# Fate of The Lost Luna

*The Betrayal of Alpha, A Luna's Return*

**Adegboye A. O.**

Fate of The Lost Luna     1

*The Betrayal of Alpha, A Luna's Return*     1

## Prologue     5

A Lone Wolf's Return     5

1     7

## Trapped     7

2     11

## Shattered Facades     11

3     17

## Stumble     17

4     23

## Bitter Brew     23

5     29

## Bound     29

6     34

## Unveiling     34

7     41

## Depart     41

8     48

## Beginning     48

9     54

## Wilderness     54

10     60

## Healing     60

11     66

Ally 66

12 72

Effort 72

13 77

Trouble 77

14 84

Ignition 84

15 90

Doubt 90

Epilogue 96

A Storm on the Horizon 96

*About* the *Author* 98

Acknowledgments 99

# Prologue

*A Lone Wolf's Return*

Under the pale glow of a full moon, the woodland remained motionless. As if they were alive, the shadows created by the old trees twisted and swirled, but there was one shadow that seemed to be moving with intent. A lone wolf named Luna, whose eyes were as dark as midnight, stalked the perimeter of the Northern Pack's domain, her heart burdened by a connection both familiar and distant.

She had no desire to come back here. The number of places she had temporarily called home—even if only for a night—had become a blur after years of running, of surviving alone, and of traversing strange territories. Still, she found herself drawn back to this spot for some reason. Maybe it was because she felt obligated to, or because she wanted to face the demons from her past, or because, at her core, she had faith in the pack that had been her family.

For the first time in years, she stepped foot in the Northern Pack's territory, where the scent of damp dirt and pine-filled her lungs, grounding her. She was flooded with memories of the things she had left behind, including the comfort of a shared fire, the power of solidarity, and the security of belonging. However, beneath those reassuring recollections lay darker ones—recollections of fights fought, of wolves lost, of a life shattered by secrets and betrayal.

Nobody knew anything about Luna's past, not even her. Still, she had arrived at this very moment, her wounded but unbroken heart and her wild but cautious soul. She silently vowed to herself that this time would be different as she returned to the house that had once been her home, feeling the burden of her past rest on her shoulders.

On this occasion, she would establish her position within the pack, leaving a legacy that was defined by her might rather than her wounds.

Ronan, an enigmatic figure, waited at the forest's edge, its light and shadow intertwined. He was someone she had been through it all with—a staunch ally, a ferocious protector, and maybe even her worst hurt. He was a continual reminder of what was and what might be, like the dawning of a new day. The love, devotion, and loss that bound them together remained strong even after all these years apart.

As they drew near, their eyes locking in the moonlight, Luna felt her heart race. She had attempted to escape the shadows of her past, but she knew that going back to the pack would force her to confront them. She had changed into someone other than the wolf that had run away years before. She had overcome adversity and emerged stronger than before, determined to regain the life she had lost.
With Ronan by her side and the future stretching out before her, Luna would quickly discover that her comeback was but the start.

She had always considered the Northern Pack her family. However, this was now her fate.

# 1

# Trapped

**LUNA**

Shadows crept down the walls of the corridor, reflecting the perpetual terror that trailed me everywhere as it seemed like an infinite labyrinth. Leaning against the chilly stone, I listened to the distant sound of conversation coming from the kitchen up ahead. I had to be very strategic with my timing if I wanted to get away unspotted. With only one slip-up, I would be apprehended.

As I peered around the bend, I noticed the teens congregating like hungry predators. My nerves were on the verge of betraying me as my heart hammered in my chest, but I mustered the courage to move. Praying that my footsteps would be unheard, I took a deep breath and sprinted across the open doorway. A rustle from my garments resounded more loudly than it ought to have the second my foot hit the ground.

I froze in a fit of panic. In the blink of an eye, a pair of shiny black shoes materialized in front of me. I forced myself to glance up and meet Sophie's icy stare as my dread levels rose. As if she had been anticipating this very second, her lips curled into a severe grin.

"Um, what exactly are we dealing with here?" With a scowl, she tossed her crimson locks over her shoulder. Confident, stunning, and ruthless—the epitome of a werewolf woman—she stood tall. Sophie understood that I was not everything she claimed to be. Even worse, she relished the opportunity to bring it up again and again.

"I-I simply desired to reach my room," I stuttered, fully aware that my words would be meaningless.

The sound of Sophie's laughter sliced through the air with a knife's edge. Ugh, my stammer has returned. A victim of her own shattered words, poor Luna.

The group of her companions who had gathered behind her shared a snicker. As Sophie moved in closer, I felt a twinge of pain as her fingers twisted into my hair and tugged gently. Her breath burned my cheek as she pushed in closer. "Do you really believe it's possible to slip past people unnoticed?" Very bad.

I clenched my jaw and refused to utter a word as her hold became increasingly firm. Their laughter was more humiliating than the agony. I was their punching bag and source of fun day in and day out. They tracked me down and dragged me back into this hellish situation regardless of how much I fought to escape.

Sophie released my grip, and I attempted to free myself, but her foot suddenly flew out, causing me to trip. My knees ached from the quiet thump as I slammed onto the floor. My cheeks turned red with shame as I stomped to my feet, and the sound of more laughter resounded from behind.

"You should've remained hidden," Sophie taunted, her voice betraying her sincere sentiments of sympathy. "Please, hurry up and go before I decide to have a good time."

Slamming the door behind me, I staggered back to my room, remaining silent the whole time. As I fell onto the bed, my breath came out in gasps, and the tears that I had been suppressing finally flowed. You couldn't call this a horrible day. Insomnia, relentless brutality, this was my existence. And I couldn't change the situation either.

A gentle buzz from my alarm roused me from a slumber in the wee hours of the morning. After drying my eyes from the night before, I dragged myself out of bed. At this hour, the world was peaceful; the critical members of the pack were still slumbering, and I could finally breathe.

In a rush, I threw on some baggy clothing to cover up my many wounds, both old and new. Covering up and acting like I wasn't broken was simpler than facing my feelings head-on. As soon as I stepped outdoors, the refreshing breeze welcomed me, and for a fleeting minute, I convinced myself that I could break free from my existence.

Just as the sun was about to set, it spread a gentle pinkish-orange tinge across the sky. This was the one moment in my life when I knew I could do anything I wanted. The outsider who hadn't changed and didn't fit in was ignored. The stranger who, long ago, had been discovered abandoned in a river.

I couldn't recall my origins or the group from which I had initially emerged. The agony was my only reality. The anguish of being unloved, of never being accepted, of never becoming the wolf my destiny called for.

I strolled into the forest, feeling the slight scars on my arms from a fire that I hardly recalled. I had no idea what had happened in the past, and the burns only added to my confusion. I felt the agony like a second skin; it was genuine, though.

I was concerned, even though nobody else was. What mattered to me was the life I may have led, the wolf that I was meant to be. Finding my place in the world became less of a goal with each passing day. All that remained was this: this solitary stroll through the forest, where I could temporarily imagine myself as part of something bigger.

I knew the tranquility was fleeting, so I made my way back to the packhouse as the sun rose higher in the sky. The others would eventually awaken, and the nightmare would resume. Surviving another day of suffering and feeling worthless was all that was left for me to do.

However, I sensed in my heart that something needed to shift. The reason is that I can't continue living in this state indefinitely.

Although the literary style has been updated, the essential themes of bullying, isolation, and Luna's desire for more remain. Please inform me if you would like more chapters or if you would like extra adjustments.

# 2

## Shattered Facades

**LUNA**

My alarm went off, waking me up from the sleepless nightmares that had been clinging to me. In my cramped chamber, I stared up at the ceiling and blinked clumsily. Another excuse for the pack to get together and make me feel even more out of place was Rosie's birthday, which was today. It was her eighteenth birthday. Therefore, it wasn't like any old birthday. At this point in their lives, wolves are expected to find a partner and be welcomed into the pack with open arms.

Still, here I was, twenty years old and still... unemployed, single, and unattached.

With a heavy heart, I forced myself out of bed and made my way to the loo to wash away the sleep residue and the hidden bruises. I snapped out of my slumber when the icy water pounded my face. With a frown on my face, I peered at my reflection in the mirror.

I looked like a total wreck in my reflection, with puffy eyes from sobbing and a new bruise on my jaw from yesterday's incident with Sophie. I painstakingly put on makeup to cover up the harm, but I knew it wouldn't last. Because the scars went more profound than the skin, no amount of powder could hide them.

I wore the same casual clothes I typically did and wore my hair in a loose bun. Like I hid the reality of my identity, they served as a barrier, concealing my body's curves and edges from scrutiny. But that would be irrelevant today. I would hide in the kitchen, far from the celebration and the prying eyes.

I had hoped, at least.

A ---

The kitchen was a hive of activity as people got ready for the party. My mum, who was one of the chief cooks, was yelling out instructions and coordinating the omegas perfectly. In an effort to blend in and avoid drawing undue attention to myself, I chopped vegetables while keeping my head down.

Excessive decorations adorned the walls of the enormous packhouse, making it an ideal venue for the grand occasion. As the daughter of the alpha, Rosie might expect an absolutely fantastic birthday celebration. A large number of males on mating missions were among the guests visiting from various packs. I felt an itch of jealousy just thinking about it. The idea of finding a life partner seemed like an unreachable fantasy to someone like myself.

As my eyelids twitched in time with my heart rate, I drew a brief breath and wiped the sweat from my forehead. I couldn't stop now, even though the makeup was starting to smudge. Along with breakfast and lunch, we had to get everything ready for tonight's feast and make drinks for all the visitors.

As she swiftly passed me, my mum tapped me on the shoulder. "No problems at all, Luna. Well done.

I put on a brave front, even though it was all an act. Dissatisfaction ensued. I needed to blend in. I had to do that if I wanted to stay alive in this dangerous place.

The sun was beginning its descent below the horizon when the celebration began, casting a warm golden light over the packhouse. Staying in the kitchen was my strategy to escape the worst of it, but alas, destiny had other ideas. Someone asked me to get them beverages, and I was called upon.

In an effort to calm myself, I pulled the black apron over my dress and adjusted it. I was taking a risk by serving at this kind of function. Wolves would be staring at me, and they wouldn't think twice about picking apart my shortcomings or, worse, venting their anger at me. No, I really didn't have a say in the matter.

I entered the majestic hall while holding a tray of champagne glasses. There was music, laughter, and the clinking of glasses filling the room. Compared to my humble abode, this was an incredibly privileged world. I skilfully maneuvered through the crowd, presenting champagne to the members of the pack, all the while being cautious not to make eye contact for too long.

I was practically unnoticeable.

I was interrupted from my meticulous regimen by a voice that beckoned to me. "Champagne?" I was signaled to by a towering wolf with sandy locks and intense blue eyes. Even though he wasn't an alpha, his mere presence gave off an air of authority. His demeanor betrayed that of a gamma or a distinguished warrior.

While trying not to lose my cool, I stutter-spoken, "Y-yes, sir," and offered him a drink. He kept staring at me for an extra second, and I drooped my head in the hopes that he wouldn't press the issue anymore.

A sudden, acute pinch appeared at my side as I turned to depart. Suddenly, I felt a lump in my throat, and the tray slid from my fingers. Broken glasses crashed to the floor, causing champagne to splash across the shiny hardwood. Whispers and suppressed laughter followed a shared gasp that reverberated across the audience.

My heart dropped as I beheld the devastation I had caused. A neighboring wolf's fitted suit was irreparably stained by the golden liquid that had poured across it. He shifted his gaze to meet mine, his countenance growing gloomier. My voice hardly rose above a whisper as I hurried to express my apologies.

"I-I'm so sorry..."

He dismissed me before I could complete my sentence. "Ignore it. Take a moment to tidy up.

I got down on my knees and gathered the shattered glass with trembling hands, and tears were on the verge of falling from my eyes. Blood spurted from a little gash on my palm as my fingers scraped against jagged edges. I grimaced in pain. No, I persisted. Just not feasible for me.

Just as I was about to finish, I sensed someone standing behind me. Steven, the son of the alpha, stood above me, his countenance opaque as I looked up.

His voice was noticeably lower than I had anticipated when he enquired, "Are you alright?"

I kept my head down and nodded. "Yes, Alpha."

He remained silent for a while, observing me as I placed the final glass on the tray. After finally saying, "Be more careful next time," he turned away and didn't say anything else.

I looked across the room and saw Sophie standing there, smirking, as I stood there with the tray in my arms. Her contentment was contagious as she muttered something to her pals, and I could almost feel it. Her goal in wanting to see me embarrassed was just this.

With my heart racing, I hastened back to the kitchen. I placed the tray on the floor and leaned against the wall to steady my breathing as soon as I arrived. The humiliation I felt within was much worse than the agony in my hand.

It was my second failure. It was always there, a reminder that I didn't fit in, and I could never get rid of it.

A ---

As the night wore on, I did my best to stay away from the main hall. When the clock struck midnight, I was completely drained. I had averted any other catastrophes, but the knot in my body wouldn't budge. Even after removing the last drop of champagne off the floor, my hands trembled.

I was worried that I wouldn't be able to continue for much longer. Continually disguising oneself as if nothing ever wronged, day after day.

However, I had a sneaking suspicion that something would eventually give out, whether it was my delicate self or the universe I had constructed to shield myself.

# 3

## Stumble

**LUNA**

The following morning arrived much too soon. The unseen weight of yesterday's setbacks made my body feel heavier than usual. Last night's humiliation remained a sharp and fresh ache, no matter how much I tried to tell myself it didn't matter. The sight of Steven's frigid stare froze me, and the sound of Sophie's mocking laugh resounded in my mind whenever I considered the broken champagne glasses.

Coughing up the will to get out of bed, I moaned. Even though it was a fresh day, being a part of this pack felt like living in a never-ending loop of slavery, terror, and waiting. Holding out hope that things will improve, even if I had lost any clue as to what that may entail.

I could still see the slight mark on my cheek where Sophie slapped me. I put on a full face of makeup in the vain hope that nobody would see it. Although cosmetics had their limitations. No matter how diligently you attempted to conceal the truth, it always seemed to find a way to surface.

A ---

As soon as I emerged from my chamber, the packhouse was teeming with activity. There was a distinct stench of lingering food and half-empty drinks all around the place as a result of Rosie's birthday celebration. The higher-ranking wolves were engrossed in their breakfast, completely unaware of the mayhem unfolding around them, as omegas frantically cleaned up the debris.

While avoiding the other wolves, I crept silently towards the dining hall. At least for the time being, I was able to escape the scowls and muttering comments because I wasn't assigned breakfast duty today. I knew better than to assume I could remain undetected indefinitely, even with the temporary relief.

A recognizable voice interrupted me as I was about to slip past the kitchen.

"Luna."

With a grip on the doorknob, I went absolutely still. With a deliberate pace, I pivoted to confront Sophie, who stood with her arms crossed and a smug expression on her face. I stood out in my ragged pants and baggy hoodie next to her perfectly coiffed hairstyle and spotless outfit.

"Good morning," I said, already preparing myself for the insult she was sure to have in store.

While approaching, Sophie drew nearer, her shoes making a loud clicking sound. It appears like you're becoming accustomed to being irresponsible. After the bash, you're back to your ghostly ways. Has your sense of purpose wholly faded?

I forced myself to remain composed by swallowing firmly. "I'm just trying to stay out of the way."

"Stay out of the way?" The sound of her savage laugh resounded through the room. What a sweetheart. So you believe that by remaining hidden, we will overlook your complete insignificance?

I held my mouth because I knew responding would make matters worse, even though her remarks cut deep. The futility of arguing with Sophie was something I had realized a long time ago. She took great pleasure in authority and reveled in your downfall the more she witnessed it.

Under my breath, I murmured, "I'll try harder," as I looked down at the floor.

"Good," she said, her voice betraying her sincere desire to be sweet. "I would hate for you to embarrass the pack again."

Right then, she pivoted and strutted away, leaving me standing there with my heart racing. My thoughts raced with images of her nasty smile as I remained motionless until I knew she had left the dining hall. Then I snuck out of there as fast as lightning.

A ---

With my hands occupied and my mind elsewhere, I continued to clean throughout the remainder of the morning. After the party, the pain in my chest intensified, and I still couldn't get rid of the nagging feeling of loneliness. Everything, not only Sophie's remarks, lingered in my mind. I felt like an outsider in the herd because of how they treated me. This was the way I was stuck. How I felt like an outsider, never entirely fitting in and always wanting something I couldn't put my finger on.

When the sun was at its highest point, I was already lost in thought as I meandered into the forest, following the same old route I used whenever I wanted a break. There was complete stillness in the woods as the towering trees swayed softly in the wind. The peace of the forest allowed me to catch my breath. There, I wasn't the target of the pack's wrath. All I was was myself.

The calming sound of the creek, as it went past the rocks and bubbled, helped calm my anxious nerves as I walked downstream. What I really want is to be able to keep on walking and let myself get carried away by the river to a place where no one recognizes or cares about me.

However, it was all in my head. Like the scars on my body, I knew it.


A ---

I returned to the packhouse just as the sun was setting, sending long shadows across the courtyard after an afternoon that seemed to fly past. I felt physically exhausted and mentally burdened by the day's events. I so wanted to go to sleep and let go of everything, but I knew that wasn't going to happen.

Steven stepped out, his scowl visible on his face, as I neared the entryway. I attempted to evade him, but he saw me in the blink of an eye.

He said, "Luna," in a harsh voice. "A word?"

With a heavy heart, I nodded and retreated to the packhouse. I was never comfortable being around him, even though he was never as obviously nasty as Sophie. A distant and icy air enveloped Steven. His presence alone was sufficient to degrade me; he needed no insults to do so.

He started by saying, "I heard about the party," while staring directly into my eyes. "And I'm not just talking about the spilled champagne."

It crushed me. He was well-informed about the incident. Gossip was the lifeblood of the pack, and they would magnify each error I committed.

I hastily remarked, "It won't happen again," in an effort to appease him.

Steven's expression changed, but he still scowled. I am not seeking remorse. It's just that I'm worried. No one else is like you, Luna. "You have never been."

Even though I knew he wasn't intentionally hurting me, his remarks hurt. It was the truth, and I had spent years evading it. Unlike everyone else, I stood out. Definitely not me.

"Just be careful," he finally remarked, his voice becoming more kind. "This pack isn't kind to outsiders."

He then walked away, leaving me to stand alone in the deserted corridor, his words still echoing in my head. The pack's cruel nature was clearly evident to me. The reality felt considerably harder after hearing it from Steven.

As I walked to my room, I rubbed the bruise on my cheek and moaned. I was emotionally and physically exhausted after the day. I required a break. I drew the covers closer about me as I climbed into bed, but I still knew it would be hard to fall asleep.

The truth was always there, lurking in the quiet, regardless of my efforts to ignore it. I felt out of place here. That was never me.

# 4

# Bitter Brew

**LUNA**

As I stood in the packhouse kitchen early the following morning, the scent of freshly brewed coffee permeated the air, spicy and bitter. Whatever the case may be, regularity has always been a source of solace for me. I was able to divert my attention from the persistent worry that had taken home in my chest by focusing on the mundane chores of grinding the beans, boiling water, and making breakfast.

Gazing at the steam rising from the pot, I dipped my head over the counter. Even though I was fidgeting, my mind was racing. The words that Steven spoke to me last night continued to reverberate in my mind. Unlike everyone else, I stood out. Definitely not me.

It was really perplexing to me. Is it a caution? An ounce of compassion? Is it more than a simple admission of truth? It made no difference; I was different, nevertheless. I felt like an outsider even in my hometown, where I had lived for many years.

The impending sound of approaching footfall jolted me out of my reverie. As soon as I realized someone had entered the kitchen, I stood up straight and turned to confront them. It was only my mum; I was relieved; her face bore the same weary look as it did every morning.

In a low voice, she saw that I had been up for quite some time and went over to assist me with making breakfast.

I was too nervous to say anything, so I just nodded. After all she had done for me, the last thing I wanted to do was burden her with my concerns. After the pack took me in, my parents were the only people who could brighten my day and show me genuine concern. The brutality of pack life was too much for even them to bear.

I managed to get out, "Steven spoke to me yesterday," but my voice was hardly audible.

In a worried look, my mom glanced over at me. "What was his statement?"

No, it's just that I stand out from the crowd. "Be careful," he warned me.

A trace of anxiety crossed her face, and her countenance hardened. As she sighed heavily, she used a towel to wipe her hands. Even though he knows the truth, Steven is a kind man. You are not treated fairly by this pack. Our best efforts have been made to protect you, but..."

"But it's not enough," I concluded for her, feeling the weight of the reality sink in.

Her quiet conveyed more than a word could have. The reality that I was an outsider would remain unchanged regardless of my parent's best efforts to shield me. I wasn't one of those real wolves. I hadn't moved one inch. I was still lost.

Undoubtedly not.

A ---

As the morning progressed, I occupied myself with routine chores, attempting to evade the watchful gazes of the other pack members who dipped in and out of the kitchen. While some of them were courteous enough to disregard my presence, others... weren't quite that kind.

With voices barely audible but distinct enough to pick up on fragments of conversation, I overheard them whispering as they went by.

"It hasn't changed yet?"

"She's fairly human-like."

"Why is she on this planet?"

I had adapted to the slap-like impact of each phrase. Their vicious taunts and insults had hardened me over the years. I had no place in their universe. All they did was put up with me. Very little.

Things in the kitchen were starting to heat up by mid-morning. The morning meal had passed, and now it was time to get ready for yet another day of seemingly unending tasks. I had gotten good at blending in with the house, cleaning, and organizing without attracting attention to myself, and the pack was expecting nothing less than perfection from me.

That was the first strategy, however.

A ---

I could hear his voice as I wiped off the counters.

I said, "Luna."

My stomach was in knots, and I went limp. Steven stood in the doorway, staring directly at me as I slowly turned to face him. On this occasion, his expression was distinct— more subdued, almost worried.

After indicating that he needed to speak with me, he beckoned for me to come after him.

For a split second, I pondered his demands, but I couldn't bring myself to defy him. My heart was racing as I silently trailed behind him as he left the kitchen and made his way down the hallway.

In his modest but tastefully appointed office, he showed me the ropes. The space was brimming with books and documents. I could feel the cool air on my face and detect a subtle aroma emanating from the shelves adorned with leather-bound volumes. With a contemplative expression on his face, Steven examined me from behind his desk, leaning against it.

"I've been reflecting on my previous statement," he started, his voice composed yet soft. "And I think it's important for you to grasp something."

With a slight nod, I waited for him to continue, swallowing my words.

"Lunara, you are unique. That is plainly apparent. But that's no indication of your worthlessness.

His remarks caught me off guard, and I briefly doubted that I had understood him accurately. Who or what is valuable to me?

He went on to say, "I know life here has been... difficult for you," his expression changing as he spoke. But please understand that you are not in this alone. Ensuring your safety is my top priority.

Unsure of what to say, I blinked. Cold and aloof has always been Steven's personality. Now, though, his remarks had an honesty and warmth that took me by surprise. At first, I didn't get it.

A scarcely audible "thank you" escaped my lips.

With an air of seriousness, Steven nodded. Things may soon improve for you after all you've been through. All you have to do is wait.

I felt a glimmer of optimism ignite within me as his words lingered in the space between us. I permitted myself to think that things could improve for the first time in a long time.

A ---

A flurry of activity ensued for the remainder of the day. With Steven's comments racing through my head, I got back to my chores. I had no idea what he meant by "change," but just the thought of things maybe turning around for the better kept me going.

By the time the sun had set, I was back in the kitchen, finishing up the remnants of the pack's lunch. At this point in the day, the room's background noise had subsided to a low hum. The sound of a creaking door opened behind me as I was wiping off the counters.

Alpha Titan's imposing presence loomed over the room as I spun around to see him standing at the doorway. His dark eyes were menacing and analytical, and he towered over everyone.  As my breath became lodged in my throat, I quickly lowered my eyes in order to avoid making gaze with the person.
Whispering "Alpha," I bowed low in reverence.

At first, he remained silent and stared at me with such intensity that it pricked my flesh. The weight of his oppressive stare was palpable. For a split second, I considered the possibility that I had upset him in some way by doing something wrong.

"You make good coffee," he finally remarked, his voice low and authoritative.

I glanced up in astonishment at the compliment, which had taken me by surprise. It was no secret that Alpha Titan did not often shower compliments, particularly on someone like me.

"Th-thank you," I stutter-spoken, for want of better words.

His steps resounded as he turned and left the room, nodding once more. My mind raced, and I froze for a second, trying to make sense of what had transpired. Even though it was just a few words, it meant a lot coming from someone of his caliber.

Perhaps, if only for a moment, things were starting to shift.

# 5

# Bound

**LUNA**

Today, I could feel the packhouse walls pressing in on me from all directions. The air was thicker than usual. Something was nagging at me, like something was ready to pop out from under the surface. I felt a subtle unease as if something was about to shift, but I couldn't put my finger on when or how it would happen.

I slipped silently along the corridors, hoping no one would notice me. Even though I had errands to run, I wasn't paying attention to them. My mind instead wandered to the discussion with Steven, the peculiar encounter with Alpha Titan, and the surprising praise. Except for the people who actively worked to bring me misery, I had gone unnoticed for all these years. So, what gives? Why had I suddenly been their focus?

It was really perplexing to me.

I was on edge when I got to the common area. Dishes clinking and soft chatter filled the air as the omegas finished off breakfast. I sneaked past them and made my way to the laundry room, which was hidden in the rear of the home.

I had only gone a short distance down the hall when someone swung open the door to the office, and I almost ran into them as they stepped out.

"Pay attention to it," an unseen voice said.

As I peered up into those familiar, icy eyes, I staggered backward, my breath caught in my throat. The legendary titan is known as Alpha. I was shivering just being in his presence; he hadn't touched me once. That wasn't necessary. He exuded an aura of immense power that rendered me feel diminutive and unimportant.

"S-sorry," I whispered, my voice hardly audible.

His expressionless dark eyes swept over me as he stared at me for a while. I was prepared for a stern warning or, worse, an outright rejection. He remained silently observing me as if he were perplexed by something I couldn't fathom.

Suddenly, it transpired.

Something changed all around me. It was as if something deep inside me had been reawakened for the first time; a peculiar warmth spread over my body and settled into my chest. My heart was racing, and I was gasping for air. It was as if the atmosphere between us had grown heavier; I could feel a magnetic pull bringing me closer to him.

My partner, "M-mate..." The scarcely audible word escaped my lips before I had a chance to stop it.

The recognition flickered in Alpha Titan's eyes as they flashed. His dominant and fierce wolf briefly sprang to the surface. Subsequently, he returned, his demeanor becoming more rigid, yet his posture had changed.

It was evident to him.

I couldn't unsee or feel the truth, which was like a spark lighting in the darkness. This man, this menacing, unstoppable Alpha... belonged to me. I became his.

However, he remained silent.

With his gaze fixed on mine, he stood eerily over me. The strain was nearly intolerable, and the silence was ear-splitting. My intellect begged me to turn away, to escape his stare, but I couldn't. My heart was pounding. The longer I remained there, the more the realization dawned on me; it was as if I were shackled, pinned by an unseen power.

Ronan, the Alpha Titan, was my partner.

I didn't understand the meaning of the word. Throughout my life, I've hoped to find my soulmate, but I never thought it would be this difficult. He is not the type of person I expected.

"Alpha —)" I tried to start, but the pressure of the moment caused my voice to crack.

Just as I was about to continue speaking, he pivoted and strode away, his footsteps resonating down the corridor as though nothing had transpired. I just stood there, motionless, staring after him, my thoughts racing in all ways.

It was evident to him. Additionally, he had departed.

A ---

Everything that followed was a haze. I was so distracted that I couldn't do a single task at home. In the blink of an eye, my entire world was turned upside down, and I was at a loss for what to do next. I enjoyed the company of a companion. My partner was the alpha of a formidable pack.

However, I was left wondering what the implications were for myself.

A part of me had always hoped for this day—the day I would meet my soulmate, the person who would put everything right in my life. But now that the time had arrived, my emotions were consumed by terror.

It was risky to be in Ronan. Word on the street was that he was a vicious wolf who would not hesitate to kill his enemies. Plus, he was now legally obligated to me.

Still, he hadn't laid claim to me. Aside from that little exchange in the hallway, he had paid me no mind at all. The unpredictability was the most frustrating aspect. I was confused about his intentions and had no idea what to anticipate. Would he turn me down? Had it been his motivation for leaving?

A chill of terror coursed through me at the very suggestion. The stories of mates being rejected and put aside, as if the link didn't matter, were something I had heard. Even more terrifying than the prospect of being bound to someone like Ronan was the prospect of his rejection, of being seen as nothing more than a possession. A ---

I remained in the shadows that night, my heart burdened by the events of the day as the pack assembled for supper. It was so hard for me that I resisted eating and talking. Regardless of our desires, we were suddenly inseparable, and all I could think about was him.

I looked up from the table to see Sophie laughing with a group of her friends when her voice cut through the background noise. She exuded an irresistible energy, her beauty mirroring the sharpness of her remarks. I could feel it coming on; she hadn't acknowledged me just yet. Especially tonight, I didn't feel like getting into another argument.

With my mind racing, I quietly left the dining hall and returned to my room. It was believed that the relationship between partners was holy and unbreakable. In contrast, Ronan made it feel like a curse.

A ---

My thoughts wouldn't stop racing later that night as I rested in bed, gazing at the ceiling. His strong posture, intense stare, and black eyes flooded my mind every time I shut them. The meaning was lost on me. It was all really confusing.

As the lowest-ranking member of the pack, I was an omega. How could I be betrothed to an Alpha like Ronan? Everything I had ever known was challenged by it. Still, here I was, entangled with him in an uncontrollable force.

But how did that impact our situation? What about me?

I knew one thing for sure as the night dragged on and sleep remained elusive: nothing would ever be the same again.

# 6

# Unveiling

**LUNA**

The following morning, my stomach felt like it was going to burst. Everything that happened yesterday in the hallway—the mate bond's irresistible pull, Ronan's heartless dismissal—kept coming back to me. I wanted to pretend it was all in my head and that it was just a mistake, but I knew in my heart that it was real.

We were partners.

From the time I was a little girl, I'd heard tales of wolves forming a mating bond with the individual meant to round out their pack. I wasn't expecting it to be so powerful; it ate up my every waking moment and left no space for anything else.

Still, Ronan had ignored me. Absolutely not. He had disengaged as if the tie were irrelevant. I still didn't know what was going to happen next, and my heart was racing with a mix of fear and optimism.

Anxieties set in.

A ---

Throughout the day, I made an effort to immerse myself in my responsibilities, thinking that the monotony would take my mind off of the emotional turmoil that was building inside. However, my pulse would quicken, and I would fret over whether or not he was close by whenever I caught sight of one of Ronan's pack members. It would make me wonder if he would finally confront me today. Choose to ignore me or turn me down.

A shiver went down my back at the mere idea. If he turned me down, what would happen? Maybe I wouldn't make it. Mates had an intense relationship, and the pain of rejection was believed to be more profound than any bodily harm. I didn't think I could manage that level of emotional, soul-crushing anguish after everything I had been through.

Quickly, rumors began to circulate among the pack, suggesting that Ronan's pack had arrived to supervise a significant matter—the nature of which remained a mystery to everyone. After breakfast, I stayed out of the kitchen. However, the omegas I passed whispered about the Blood Moon wolves, how they were so terrifying, and how they had put a damper on the entire packhouse.

Still, they were clueless. The mate bond's secret, which was burning within me, was unknown to anybody. The truth was so horrible that I doubted I could face it.

A ---

My peaceful afternoon was shattered at that hour. As my thoughts wandered to Ronan and my future, I was washing the foyer's enormous windows when I heard a voice calling out to me.

I said, "Luna."

The sound was immediately recognizable, and I tensed up in response. My breath caught in my throat as I cautiously turned to see Alpha Titan standing in the doorway. He stared at me with deep, mysterious eyes.

We remained motionless, gazing at one another for an extended period. I resisted the need to look away as my heart pounded in my chest. Like the first time our mate bond surged, his presence was overwhelming. However, at this point, my thinking was apparent.

"M-Mate," I muttered, my voice quivering. Words flew out of my mouth before I could stop them, even though I didn't intend to.

Ronan maintained her expression. With his eyes fixed on me the whole time, he cautiously advanced. It seemed as if we were precariously perched on the brink of some dangerous event; the anxiety was heavy and tangible.

I thought you might be familiar with this. A chill ran down my back as I listened to his deep, nearly growling voice.

My mouth was parched, and I forced myself to swallow. So, yeah. You're my partner.

Even though they were the truth, the words tasted funny when I tried to say them. I couldn't continue to deny it, no matter how terrible it was to confess.

Ronan drew nearer, drawing us closer until we were standing within a foot of each other. I could sense the intense heat and sheer force that radiated off of him. However, there was an additional element—another, more sinister, and potentially harmful one.

"I can feel the bond," he remarked, his voice betraying his disappointment. "Yet such a thing is impossible."

I felt as if the force of his remarks had sucked my breath out. I attempted to make sense of what he was saying as I gazed at him with wide eyes. Does this really have to take place?

"I am the leader of the Blood Moon pack," he added, his expression becoming more serious. "Keep the interruptions to a minimum. "Complicated things are not what I need."

Will there be complications? I was surprised by how feeble my voice came out; it was hardly audible. "Are you implying that I'm an inconvenience?"

Anger or sorrow could have been the emotion flashing through his eyes. Whatever it was, though, it vanished just as fast as it had emerged. With his hands clasped tightly at his sides, he tightened his jaw. "This bond... doesn't work for me," I'm saying.

When he spoke those words, my heart broke. He wasn't explicitly rejecting me, but his comments conveyed a strong message. He had no interest in this. He was entirely against my being in his life.

It was worse than I had imagined; it was like someone had twisted a knife into my chest and forced it into my body. After years of being made fun of and told I didn't matter, even my partner, who was meant to treat me with respect, was ignoring me and even trying to pull me away.

"Excuse me, but I'm confused," Tears hurting my eyes, I choked out. I asked, "Why?"

There was a fleeting easing of Ronan's expression. You have no idea how heavy my burden is, so you can't possibly comprehend. A pack that doesn't value you recognizes you as an omega. You've always been one step ahead of the pack. I, on the other hand, must not be feeble. I feel exposed because of this connection.

My eyes were on the verge of falling as I forced myself to swallow. "You are not weaker because of the link. We're told it will fortify us. Our purpose is to —

Just stop. He interrupted me abruptly with his harsh voice. He moved away from me, taking a step back. A vulnerability like this is something I cannot afford. Currently, no. Absolutely not.

I nervously chewed on my lip, attempting to contain the avalanche of feelings that were about to consume me. Everything I had worked for seemed to be slipping away as if the world were collapsing around me.

"So, after this?" In a calm tone, I enquired.

Once again, Ronan's piercing gaze fixed on me. "We don't tell anyone about this. It is a mystery. And for the time being... not much changes.

The situation remains unchanged. The words resounded in my head, desolate and lifeless. Although he didn't explicitly reject me, I felt rejected nonetheless.

My pulse was pounding in rebellion as I slowly nodded. "Alright," I mumbled, despite the fact that it was the very last thing I desired to utter.

Standing there alone under the crushing weight of his decision, Ronan turned and walked out of the room without saying a word.

A ---

Everything that followed was a haze. My thoughts were elsewhere while I merely performed the tasks at hand, such as cleaning and assisting in the kitchen. My mind kept wandering to Ronan, the bond, and the agony that was now writhing in my chest.

There was an expectation that the mate bond would be a priceless present. But suddenly, it seemed like a curse, a constant reminder of my shortcomings. He didn't think I was strong enough, good enough, or worthy enough. Despite my best efforts to deny it, the reality was right in front of me.

Even my mate despised me because I was an omega, doomed to a life of shadows.

A ---

My heart was aching from the weight of all that had transpired, and I rested in bed that night, gazing at the ceiling. I was at a loss for actions and directions after this.

We had a genuine connection, Ronan and I. There was an indescribable connection between us, and I could feel it throbbing under my skin. However, he had previously expressed his disapproval, viewing it as a potential flaw.

To force myself to stop crying, I shut my eyes. Abusive treatment, scorn, and rejection were just a few of the things I had overcome. However, this was distinct. I had my doubts about my ability to endure this.

For the simple reason that I had never felt more alone despite my best efforts to the contrary.

# 7

# Depart

**LUNA**

As the minutes dragged on, the hours dripped by like raindrops on a windscreen. The image of Ronan's icy, unfeeling visage lingered in my mind whenever I shut my eyes. His harsh remarks served as an ever-present reminder of how I fit inside his universe.

I felt out of place.

The packhouse was even more stuffy than before. Wolves who could feel the tension in the air whispered to each other as the heavy aura of Alpha Titan's pack hung over us. The atmosphere was electric with chatter and speculation about what might happen next. The truth that was between Ronan and me was, of course, unknown to everyone. Our mate link remained a well-guarded secret.

But I couldn't bear the burden of keeping it secret any longer.

A sense of emptiness persisted. Ronan had previously indicated that he would not flat-out reject me, but he also refused to recognize me as his partner. Waiting for something that may never come, he had abandoned me in a state of hopelessness and limbo.

This way of life is too much for me to bear. Something had to be done. In some tiny way, I had to stop letting my life happen to me.

A ---

Long shadows stretched over the floor as the early sunbeams slipped through the drapes. I had hardly slept because my mind couldn't settle down. I had the solution in my head. It was the last option I had, and I didn't much enjoy making it.

Unfortunately, I had to depart.

The relentless pressure, the continual reminder of all the things I will never be, and the packhouse itself were enough to make me want to run away. Every day, I couldn't approach Ronan and pretend our connection didn't exist because I lacked the strength to do so. My heart raced with fear at the prospect of meeting him again and understanding his feelings.

I was hoping that everything would become more evident once I departed. Perhaps the separation would shed light on my next move, on how to proceed while the connection clung to me like an unbreakable link.

A ---

My hands seemed to be operating on autopilot as I hurriedly stuffed garments into a tiny bag. Just enough time away to regroup and figure out how to make it in a world where my soulmate doesn't desire me—that's all I intended during my brief absence.

A sudden tap at the door startled me as I was zipping up the bag. I was taken aback, my pulse pounding, as I wondered who could it be. It was Ronan, I thought for a second. Unexpectedly, it wasn't he who appeared at the door.

Who was that? My mum.

"What is Luna?" She entered the room with a gentle, worried tone in her voice. I could see she was worried as her gaze shifted to the luggage on the bed. Let me know what you are up to.

To avoid seeing her eyes, I averted my gaze. "I have to leave this place for a bit."

With a wrinkled brow, she moved closer, her palm delicately brushing against my arm. "Why? What's happening? Is that the group? What just transpired?"

I was going to inform her. The anguish that was consuming me, the mating link, and Ronan were all things I wanted to tell her. However, I was unable to do it. The weight and constriction of the words made it difficult for me to speak.

"I simply... require some distance," I said. "Allow me some time to reflect."

Although my mother's expression relaxed, her worry remained etched on her face. Her expression betrayed her want to dispute and prevent me from departing, but she eventually nodded. I understand. Tell me you'll be cautious, though.

Assuring her, "I will," I wasn't sure whether I really believed it.

She gave my arm a light squeeze and then turned to exit the room, but she stopped short of the door. "Hey Luna, you know I'm always here if you ever need to chat, okay?"

As the constriction in my chest increased, I gave a little nod. I am aware.

She then departed, leaving me to face the encroaching stillness on my own.

A ---

Sneaking by the other wolves in the packhouse, I made my way through with relative silence. I was trying to avoid being the center of attention. I preferred it when no one enquired. I just wanted to escape the crushing burden that had taken root in my life and breathe again.

As I stepped outdoors, the sun was already high in the sky. The cool air filled my lungs as I made my way away from the packhouse and into the forest. A gentle breeze caused the branches of the towering trees to sway as they spread out. Whenever I needed to get away from it all and clear my thoughts, this was my go-to spot.

The familiar trail made its way through the trees, and I could feel my anxious nerves easing as I listened to the adjacent stream. I felt lighter and lighter as I walked, as if I were escaping the mayhem in my life with every stride.

However, not even in this tranquil forest setting could I evade the fact.

At the stream's brink, I knelt to let the refreshing water trickle between my fingertips. To calm myself, I closed my eyes and listened to the river. I tried to picture a world where Ronan had embraced me and where our mate link had united us rather than divided us.

However, the harsh truth constantly surfaced again and again, regardless of my best efforts to ignore it. There was no way to escape the fact that I was shackled to someone who despised me, and I was all alone.

I just sat there, thinking deeply, looking out over the lake. When I heard footsteps behind me, I had no idea how much time had gone by. I spun around, partly hoping to find Ronan standing there, and my heart skipped a beat.

But he wasn't the one.

Steven, it was.

A ---

"Luna and I could tell Steven hadn't stumbled into this place by the gentleness of his voice; it had a hardness and weight to it. "This is no way for you to vanish."

Standing before him, I clenched my jaw and swallowed forcefully. I had no intention of departing indefinitely. I required some personal space.

As he approached, Steven's expression became more gentle. "You have endured immense hardship. Okay, I see. However, escaping won't solve the problem.

Rather than meet his gaze, I drooped my head. "Anything else would be a complete loss."

Just for a split second, we remained silent. Steven seemed to be staring down at me with all his might, and when I lifted my head to look into his eyes, I noticed an unexpected quality—empathy.

He said, "You're not a lone wolf." "We are here to support you through whatever you're facing."

I yearned to put my faith in him. The secret that had been destroying me ever since I discovered Ronan was my mate was something I wanted to spill the beans on to him. However, I was unable to do it. I couldn't bring Steven or anyone else into this mess with me.

The words "I'll be fine" came out of my mouth in a whisper, but I wasn't sure they were true. "I require a brief respite."

After giving me a quick once-over, Steven nodded. Go at your own pace. However, avoid being alone. Luna, we are your pack. No matter how you slice it, we're always here for you.

Unexpectedly, his remarks struck a chord with me, and I experienced a tiny flicker of hope—my first in quite some time. Perhaps I wasn't as isolated as I had imagined.

A ---

As Steven pivoted and made his way back to the packhouse, his form vanished into the forest, and I beheld his every move. The impact of his words hung heavy on me as I stood there for what felt like an eternity.

His assessment was correct. Getting away wasn't the solution. The thought of remaining here, a part of a pack where not even my mate wanted me, was intolerable. I was clueless as to how to fix the problem, but I did know this: I couldn't continue to exist in limbo. I needed to move on, even though I had no idea what that road would entail.

Gazing downwards, I observed the stream's water trickling over the pebbles. Perhaps it was time to give in to the flow. Perhaps it was time to release my grip and follow the currents.

I also gave myself permission to hope, which was a rare occurrence for me.

# 8

# Beginning

**LUNA**

Even though I stayed in the packhouse, I felt like I had changed. I didn't know what that shift meant just yet, but I did know that I couldn't continue as before, hoping against hope that a miracle would break the cycle of misery and doubt. Unexpectedly, Steven's words had given me the feeling that maybe, just maybe, I wasn't as alone as I had assumed.

However, I was still burdened by the connection with Ronan. Despite my best efforts to put some space between us, the link remained, quietly throbbing beneath the surface. No matter how far I attempted to flee, I felt drawn back to him, as if an unseen yet unbreakable string bound us together.

The stress was becoming too much for me to bear. It would ruin me if I remained here and allowed it to fester. There needed to be a shift.

A ---

As I awoke in the stillness of the morning, I could make out the gentle rays of sunshine streaming through the open windows. The customary burden of fear associated with the dawn of a new day didn't seem to be quite so heavy. I had a new level of optimism—the first in quite some time. A little spark was all it took to motivate me to get out of bed and start making things happen.

In a flash, I changed into loose-fitting, comfortable clothing that would not restrict my movement. I would not cower in fear today. I had no intention of continuing to run. On the contrary, I intended to do something daring and experimental.

I intended to approach Ronan head-on.

I repressed the terror that the decision had sparked in me. Someone had to tell me. I wanted to understand his steadfast refusal to acknowledge the relationship and his determination to act as if it didn't exist. I was even more concerned about having a firm grasp of my position.

I could no longer bear to remain hidden as opposed to this.

A ---

As I descended the stairs of the packhouse, it was already a hive of activity. Although there was a distinct boundary between our packs, wolves from both Ronan's and mine packs were there and mixed in with one another. The Blood Moon wolves were more significant and better than the others, and they commanded attention wherever they went.

With my head lowered and my senses on high alert, I made my way through the crowd. The atmosphere was electric with anticipation; it felt like everyone was staring at the action. It wasn't simply the typical posture of having several alphas in one area; there was something further, an inexplicable quality.

But at this very moment, it made no difference. Discovering Ronan was my only concern.

He was muttering to a band of his soldiers in the far corner of the room when I caught sight of him. Even when he didn't speak a word, his towering stature and intimidating presence commanded respect and admiration. The mate bond came roaring to life the second my eyes fell upon him, and my heart raced in response.

I braced myself for what was next by taking a long breath. It was this. I couldn't go back at this point.

My nerves were jangling with a mix of fear and resolve as I walked gingerly towards him. Before I even got to him, Ronan whirled around to face me, his black eyes narrowing slightly as they met mine. He must have felt my approach.

I called out to him, "Alpha," sounding more assured than I actually was.

With a short nod, he sent his soldiers fleeing into the crowd, leaving just the two of us standing in the center of the packhouse. For a split second, we remained silent. I could feel the thick and heavy weight of the tie between us, but I also saw the stiffness in Ronan's jaw and the way he was holding back.

"Whatchamacallit, Luna?" Low and nearly growling, his voice betrayed no ill will.

I forced myself to swallow, drawing on the inner strength I had developed over the early hours of the day. A conversation is in order between us, in regards to... our own.

With his arms crossed over his chest, Ronan's face turned gloomy, and his mouth clenched. "No one represents 'us,' Luna. I mentioned that earlier.

I had anticipated his hurtful remarks. Nevertheless, I was unyielding. Despite the shaking in my hands, I managed to say, "You can't just pretend this doesn't exist.". "There is a genuine bond. Both of us are aware of it.

Ronan's eyes widened, and for an instant, I thought I caught a glimpse of something—perhaps regret or annoyance—but it vanished just as fast as it came. As he crept closer, his imposing presence began to tower over me.

"You don't comprehend your request," he stated, his voice low but tinged with menace. "I feel exposed because of this connection. Additionally, that is out of my price range. At this very moment, no.

Feeling the intensity of his closeness, I retreated a step, but I would not be intimidated. I looked him in the eye and asked, "The truth, please. Nothing else. "I am worthy of that amount."

Ronan fixed his intense gaze on me, but suddenly, there was a change in his look that I couldn't identify. His previous display of icy detachment had disappeared. Something more profound, something contradictory, was at work.

He remained silent for an extended while. Heavy and stifling was the quiet that hung between us. The strain was becoming apparent as the weight of unspoken things weighed down on us both.

At last, he broke the silence.

He spoke in a low, harsh voice, yet he was correct. "There is a genuine bond. Plus, I can't ignore it. Still, it doesn't imply I'll be able to embrace it.

I had to fight the want to freak out in response to his comments, but I kept my cool. Just say no. Why are you hell-bent on ignoring me?

For an instant, Ronan's expression softened, and I caught a glimpse of her fragility shining through her armor of control and strength. "Because I've already lost a lot," he whispered. "I simply cannot afford to lose once more."

I wasn't expecting to hear that confession. In that instant, I saw there was more to Ronan than what I had previously thought—a man who seemed unbreakable and unbeatable. An unhealed wound lay dormant beneath the surface, causing anguish.

"I'm not requesting that you give up anything," I whispered, my heart breaking for him. "Please, just let me in."

I felt for an instant that Ronan may say something more—something that would alter everything—as our eyes locked. However, his visage further hardened as he shook his head.

A simple "I can't" was all he said.

Tears started to spring up in my eyes as the gravity of his words struck me like a punch. I was hoping to find some explanation or explanation that would explain everything, so I came here. However, all I ended up with were additional inquiries.

"I get it," I murmured, my voice just audible, before turning to depart.

However, Ronan's hand suddenly extended, seized my wrist, and halted my departure. As I stared up at him, my pulse raced, and I went limp.

A low, rough voice of the word "luna" came out of his mouth. "I'm not saying no to you. "Not just yet."

There was a lot of implied meaning in the words that floated between us. I stared at him, my jaw dropping, confused about what to say and its significance.

Standing in the center of the packhouse, my mind racing and heart pounding, I was left standing as Ronan released my wrist and turned away before I could answer.

A ---

After my encounter with Ronan, I felt even more bewildered than before. I wasn't rejected, but I also wasn't accepted. There was no obvious way out; it was like being in a limbo state.

I couldn't escape the sensation that something had changed between us as I walked back to my room. Although it fell short of my expectations, a resolution was achieved. That was sufficient for the time being.

However, I had a feeling this wasn't the final chapter. There was more to this connection and Ronan than either of us had ever anticipated. I would need to be prepared for anything that occurred after this.

I mean, this was only the start.

# 9

# Wilderness

**LUNA**

I could only find oxygen in the woods.

Away from the oppressive stress of the packhouse, the tall trees and the sound of leaves rustling provided a much-needed escape. Because no one here said trash about me behind my back or gave me the creepy eye, I've always felt more at ease here, in the peaceful seclusion of nature. All I could hear was the distant sound of the stream, the wind in the trees, and the ground beneath my feet.

I needed this badly after our talk with Ronan. The weight of his comments and the unanswered questions that remained between us continued to press on my heart. Although he hadn't outright rejected me, he also hadn't fully embraced our connection. I could hardly endure the limbo he was keeping me in.

Now I'm out here, and I might forget. There was a part of me that wanted to believe things weren't as complicated from this vantage point.

A ---

I cautiously made my way deeper into the forest, keeping to the well-worn trail that meandered through the trees. The sound of gentle crackling underfoot and the refreshing aroma of damp earth settled into my lungs, bringing a sense of stability. I set off on foot without a specific goal in mind, driven only by the need to break free from the burdens that had been pressing on me.

Whenever I needed a break from the pressures of pack life, I would go to the forest. Being an omega didn't get me in trouble here. Nobody made fun of me because I didn't move. When I was in the wilderness, all anyone could see was Luna.

However, I experienced a different sensation today when I continued my exploration of the forest. Being there. The sensation of being observed was subtle yet distinct.

I remained motionless, my hearing amplified as I paid close attention to my surroundings. Birdsong, leaf rustling, and other typical forest sounds filled the air, but there was also something unfamiliar, and it made me nervous.

I was not alone out here.

My pulse was thumping as I cautiously surveyed the forest, taking deep breaths. Though I didn't perceive anything out of the ordinary, an unsettling tingle in the nape of my neck informed me that I wasn't alone.

"Who do you see?" My voice was more steady than I felt when I yelled out.

Just for a second, there was complete quiet. There was a palpable sense of anticipation in the air. A man emerged from the shadows cast by the trees and advanced towards the scene.

Seeing Steven made my heart skip a beat.

A ---

The word "Luna" came out of his mouth in a soothing but enigmatic manner. "I apologize for catching you off guard."

As I stepped back, my body relaxed a little, and I released a breath I hadn't known I was holding. The question is, "Why are you out here?"

He shrugged and came closer but yet maintained an appropriate distance. "I could pose the identical question to you."

"I needed some air," I said, shifting my focus from him to the road ahead. "There is just too much in there."

With a gentle smile, Steven nodded. My apologies.

We remained silent for a while, the awkwardness between us melting away. Whenever I was in Steven's presence, I felt an unusual tranquility, as if he were apart from the storm that engulfed me at all times. He stood out from the rest of the pack because he was dependable, kind, and stable.

Smiling, he said, "You don't have to keep running away, you know." His gaze remained set on the trees ahead. There is no one in our family but you. We are here to lend ao.

As my jaw tightened in response to his remarks, I scowled. "It isn't how it feels. I can't help but feel like an outsider when I glance around.

With his solemn stare, Steven turned to meet mine. Luna, you are not an outsider. Whether you know it or not, you're a member of this group.

I yearned to put my faith in him. I only I could convince myself that I fit in, that I was welcome here. It was challenging to cling to that hope after everything I had endured, including bullying, rejection, and the continual messages that I wasn't good enough.

"Ronan doesn't want me," I blurted out, almost like a confession, before I could control myself.

I couldn't quite make out the change in Steven's demeanor as something fluttered in his eyes. "Ronan is... intriguing."

The expression was too mild.

"He's avoiding me," I murmured. He's my partner, but he refuses to acknowledge our tie. Plus, I am at a loss as to what to do. How can I convince him that this could be successful?

For an instant, Steven didn't say a word, staring thoughtfully ahead. At last, he spoke, and his voice was gentle yet determined. If he isn't ready to see something, you can't make him. You are not involved in Ronan's motivations. All eyes are on him.

Perplexed, I scowled. I don't understand.

Steven stared me in the eyes as he conveyed what seemed like pity. "Ronan has endured immense hardship. People—important people—have departed from him. To shield himself from harm, he has constructed barriers. The relationship terrifies him. The reason is that he will have to let someone in again. And he finds that challenging.

For the first time, I began to comprehend the gravity of Steven's remarks. It was not because he didn't want me that Ronan was shoving me away. His fear of losing me was the reason he was trying to distance himself from me.

However, the pain remained unchanged.

"Then what is my next move?" My voice was low as I enquired.

Steven softly grinned. "Hold tight. Wait for him. His seeming invincibility belies his frailty. And he will see the potential in you eventually.

The measure of the unknown weighed heavily on my heart as I cast my gaze downward. Am I able to wait? Even if it seemed like everything was collapsing around me, could I maintain hope?

With my voice barely audible, I said, "I'm not sure I'm strong enough."

I wasn't expecting the comforting touch of Steven's hand on my shoulder. Luna, you have more strength than you realize. Compared to other wolves, you have already managed to survive more times. There's still time for you to succeed.

I felt a surge of courage that I hadn't experienced in a long time as a result of his words. Perhaps he had a point. Perhaps my strength was more significant than what I believed it to be.

Additionally, there is always the chance that I will discover a technique to overcome this.

A ---

There was a noticeable relief from the stress I'd been experiencing as Steven and I made our way back to the packhouse. The forest, which had previously burdened me with my doubts, now seemed lighter and more inviting. After coming here in search of a way out, I discovered something else entirely—a new point of view.

I was not alone in my continued confusion about Ronan and the precarious nature of our relationship. I had Steven, my pack, and who knows, maybe even Ronan in due time.

At the moment, all I can do is take things day by day. I permitted myself to hope that things may improve as I made my way into the woods, surrounded by the earthy and piney aromas.

That this was really only the start and not the final chapter.

# 10

# Healing

**LUNA**

The following morning, the packhouse was peaceful, as the atmosphere was heavy with the pent-up tension that had persisted for several days. It was palpable in every look and murmur as the wolves went about their business as usual, acting as though nothing had changed, but in truth, everything had.

My thoughts from my forest chat with Steven were still racing. Unexpectedly, he had shed light on Ronan, but it hadn't wholly dispelled my doubts. It just served to complicate matters further. It wasn't merely a rejection of the relationship; Ronan's emotional distancing sprang from his sorrow and dread. Regardless, I remained smack dab in the center of it all, awaiting his decision on whether or not he wanted me.

This level of waiting is too much for me. We had to make a sacrifice.

I overheard fragments of talks regarding Ronan's pack and their time here as I walked through the halls. Their anticipation of something monumental was palpable. Whatever it was, the vibe in the packhouse had changed, and I had a sneaking suspicion that whatever was on the horizon would radically alter our lives.

A ---

Around the middle of the morning, I arrived at the little clinic close to the packhouse. I didn't intend to come here, but ever since Sophie's latest attack, a dull ache in my ribs has been bugging me, so I thought it was time to take action. Although the bruise had healed, the agony was still there, serving as a continual reminder of the lengthy history of abuse I had to face.

There were very few patients in the waiting room of the clinic, so it was peaceful. I made my way to the front desk, where Dr. Maddox, the pack's physician, was engrossed in paperwork organization. A calm and mature wolf, he had silver streaks in his hair and a pair of specs resting on the bridge of his nose.

He smiled warmly and said, "Luna," as I drew near. "Why are you in this place today?"

I was at a loss for words, trying to conceal the welts that were still settling into my flesh as I explained my visit. My ribs have been giving me a little trouble recently. I figured I should have it examined even though it's probably nothing.

As he gestured for me to accompany him into one of the examination rooms, Dr. Maddox's expression softened. "Then let's take a look."

I sat down on the test table and lifted the edge of my shirt slightly to expose the blemish on my side. As Dr. Maddox scrutinized the marks, his countenance grew more serious and worried.

"How on earth did this come to pass?" he enquired, maintaining a delicate but resolve.

Unsure of what to say, I nervously chewed on my lip. Because the abuse had persisted for so long, I had become numb to it, and I didn't want Sophie to have to deal with it, either. Still, Dr. Maddox wasn't one to shrug his shoulders.

"It's nothing," I remarked hastily, attempting to downplay the situation. "I slipped and fell."

The matter was not pursued further by Dr. Maddox, who appeared unconvinced. But he just nodded and went to get a few things out of the cabinet. Even while nothing appears to be damaged, you're actually very bruised. Here is some pain medication and an ointment to speed up the healing process.

An overwhelming sense of appreciation washed over me as he worked. Never once had Dr. Maddox been overly nosy or asked questions about anything I wasn't comfortable discussing. Respect, which was rare in the packhouse, was his attitude towards me.

With a quiet "thank you," I accepted the medicine he had given me.

Even though Dr. Maddox smiled, his eyes betrayed a sense of melancholy. Luna, you can find support and help if you need it. Feel free to ask for assistance whenever you need it.

Even though I wasn't entirely convinced, I nodded anyhow. I had always struggled with asking for help. I felt like I could never reach out again after spending so much time surviving alone and covering up my wounds and pain.

Yet perhaps the moment has come to begin.

A ---

Once again, when I stepped out of the medical center, I felt myself meandering through the woods, the refreshing wind caressing my skin. As usual, the gentle swaying of the trees and the distant murmur of the stream soothed me. However, today seemed different.

I felt as if a physical burden had been heaped on my shoulders from the accumulation of recent events, including the mate bond, Ronan's rejection, and Sophie's abuse. It had been a long time since I had pretended I could manage it, that I was strong enough to bear the weight alone. In reality, though, I was exhausted. Weary of taking it, weary of being disregarded, and weary of waiting for anything, anything, to improve.

My mind wandered back to the clinic and Dr. Maddox's kind offer of assistance as I walked. Perhaps he had a point. Perhaps I could have gotten through this with some help. However, I had learned to conceal my weaknesses from an early age, so asking for assistance felt like admitting my weakness.

Still, it was appealing to think that I could rely on someone and that I wouldn't have to bear the burden of everything alone. I had previously forbidden myself from ever considering the possibility, but now... perhaps it was time to quit avoiding the thought.

It could have been time to come out of hiding.

A ---

I returned to the packhouse just as the sun was setting. A crispness in the air and a sky painted in pinks and oranges heralded the arrival of autumn. I could sense the familiar packhouse tension as I entered, although it wasn't quite as bad this time.

It had been a long time since I felt like I could make a decision. I could run away from the pain, or I could confront it directly. I had not felt this much strength in years at the prospect of taking charge of my own life, of refusing to allow Ronan's rejection or Sophie's abuse to define me. I had no idea how that would play out, but just thinking about it gave me power.

In no way was I helpless. I was strong.

You were Luna.

It was high time to begin behaving accordingly.

A ---

Later that night, as I lay in bed, basking in the moonlight streaming through the window, thoughts of Ronan returned to my mind. The connection we had was still there, throbbing under my skin, but it wasn't as stifling as it had been previously.

I hadn't permitted myself to believe in anything in a long time, but Steven's words had stuck with me and given me something to hang onto, even though I wasn't sure I could trust them.

Best wishes.

Perhaps Ronan needed more time to come to terms with the link. Perhaps he would never have been. But that didn't imply I had to do nothing while he decided. I was free to pursue my own goals and live my own life. I was also beginning to think that I could succeed even without him.

I felt the stress in my body begin to melt away as I closed my eyes and exhaled slowly and deeply. I knew there would be obstacles on the still-uncertain road ahead. Now, though, I was prepared.

I felt a lack of fear that I hadn't felt for quite some time.

# 11

## Ally

**LUNA**

The following day dawned colder than the last, the sky a washed-out gray that seemed to echo the uneasiness I still felt. But despite the chill in the air, I was determined. I refused to go back to my old ways of hiding today since yesterday had been a turning point in my life. Though I still felt the effects of Sophie's bullying and Ronan's rejection, I had started to think that I could manage them. Now, I had company.

As I made my way through the packhouse, I could hear the customary cacophony of morning activities: wolves howling, omegas rushing about to make breakfast, and the higher-ups debating the matter that had brought Ronan's pack here. It was the same as before, but I had a different feeling about it today. More powerful.

I knew I had to get something done today, but I didn't have anything specific in mind or a strategy for the day. The uncertainty was becoming too much to bear, so I had to do something about it. I needed to make a move if I was ever going to stop letting life happen to me. Re-establishing contact with the pack and with myself was essential.

A ---

I saw Molly standing by the kitchens as I made my way through the packhouse. Her countenance brightened as she saw me approaching from behind as she hurriedly stacked plates, her forehead furrowed in concentration.

"Luna!" she exclaimed, beckoning me to come over.

Among the few wolves who had ever been kind to me, Molly had always been a shining example. She showed no disdain towards my omega status or my incapacity to shift because of it. Even though I kept her at a distance at times, she remained one of my few pals.

Joining her, I uttered a soft "hi" and forced a smile. "Need help?"

Molly shook her head, but there was a glad glimmer in her eyes. "I've got it under control. You appear to be in need of a break from whatever has been troubling you.

I couldn't help but laugh softly. Without even asking, Molly could cut through the din and see through to the heart of the matter. She was discreet, but she could tell when I needed to vent.

As I leaned against the counter, I confessed, "I simply... needed to escape my thoughts for a while." "It has been quite a ride."

As her worried countenance softened, Molly shifted from stacking the dishes to facing me. "Do you mean Ronan?"

The question caught me off guard, and I hesitated for a second before nodding. "Yeah. It's... complicated."

Molly sighed, wiping her hands on her apron before crossing her arms. The way he stares at you is something I've observed. I'm not the only one who thinks this, am I?"

I let out a frustrated and sorrowful sigh as I shook my head. "No, that's wrong. However, he has been emotionally withdrawing. That bond won't budge him. Furthermore, I am at a loss as to how to proceed.

With a puzzled expression on her face, Molly scowled. So, have you given him a chance to talk? The kind of conversation with him that I mean?"

As I asserted my claim, I nervously chewed my lip. But he seems to be disconnected. He maintains his story that the bond is out of his price range and that it leaves him exposed.

Molly sighed softly as her brow furrowed. This pack requires a powerful Alpha, but I don't know anything about Ronan. And as far as I can tell, he has made a name for himself by refusing to be beaten. Perhaps... perhaps the significance of the relationship is too much for him to bear. In honor of you both.

Reminding me of something Steven had said the day prior, her remarks moved me deeply. Ronan was more worried about what the bond would do to him than it was about me. Still, that didn't help with the acceptance process.

"I know," I whispered. "Yet, comprehending that fact does not alleviate its pain."

Moving closer, Molly reached out and touched my arm. "No, that's incorrect. You underestimate your strength, Luna. You have persevered through adversity and emerged more robust than before. Hold on for the ride. Even if you're impatient with

Ronan, you know he'll come around eventually. No matter what happens, it would be best if you pressed on.

I really needed to hear what she had to say. Molly was the one who would tell me I was strong whenever I doubted my abilities. And I finally felt like I could see a path ahead today—after what felt like an eternity.

A ---

Afterward, Molly and I spent the remainder of the morning conversing about lighter topics, such as her most recent endeavors, the pack's turmoil, and the amusing eccentricities of a few of the younger wolves. It was a pleasant diversion from my problems, and for a short while, I forgot that I was an omega who couldn't shift or that I was chained to an Alpha who didn't want me.

At noon, the packhouse was more crowded than before. The sound of howling wolves filled the air as members of both packs entered and left the communal spaces. Though the animosity between Ronan and our packs remained unchanged, I felt less of an impact from it today. I finally felt like I belonged here, that I wasn't an observer on the periphery.

On my return to my room, I crossed paths with Steven in the corridor. He nodded and smiled briefly, and I felt an overwhelming feeling of appreciation towards him. His remarks in the forest had an impact on me since they offered a new viewpoint. Even though we hadn't settled our differences with Ronan, I felt better prepared to deal with the unknown.

Now, I had company. When I was down and out, I knew I could always count on my friends and supporters to cheer me on and have faith in me. And that was more than enough to motivate me and give me the fortitude to deal with whatever the future held.

A ---

As the pack assembled for dinner later that night, I was seated at a table at the room's rear with Molly and a couple of other omegas. Our conversation and meal could take place in peace, away from prying eyes. As the guests made themselves comfortable for the evening meal, the room was filled with the sound of chatter and the clink of dishes and cutlery.

As he spoke with a few of his warriors, Ronan sat at the head table, his look impossible to decipher. I peered across the room and saw him. Our connection was like a continual hum under my skin; it never went away, and just thinking about him gave me the willies. Contrary to previous times, I didn't experience a crushing despair tonight.

I had no intention of waiting for him to return. I refused to allow his choice—or lack thereof—to dictate how I lived my life. He was solely responsible for his inability to accept the relationship. Nevertheless, I refused to allow it to hinder my progress any further.

I felt a wave of calm wash over me as my focus returned to my buddies. Although it fell short of my expectations, a resolution was achieved. It was sufficient for the time being.

A ---

As I lay in bed that night, with the moonlight painting gentle shadows across the room, I permitted myself to think about the day. Even though I hadn't achieved perfection, I had made progress. I reconnected with a friend who helped me see my value in Molly, and I reached out to her. No longer was I merely existing; I was beginning to live genuinely.

I was confident in my ability to cope with whatever lay ahead, regardless of how long it took Ronan to accept the bond or what else may happen.

The reason was that I had company. Plus, I was never one.

# 12

# Effort

**LUNA**

A pattern emerged over the days, but it was less oppressive than the previous one. I decided after my conversation with Molly to stop worrying about whether or not the pack would accept me and start focussing on myself. I had finally had enough of allowing other people to determine my value.

That was no picnic. Rumors about Ronan's warriors and the enigmatic rationale of their prolonged stay continued to circulate in the packhouse, which was still tense due to his presence. The crushing weight of expectation, though, was no longer a burden. Floating aimlessly was a sensation I no longer experienced.

On the contrary, I dove headfirst into my job.

A ---

I got up before the rest of the pack the following morning. The packhouse grounds were bathed in a pale, golden light as the sun had just risen over the horizon. Wanting to get a head start on the day, I sneaked inside the kitchen. Domestic duties like cleaning, cooking, and prepping allowed me to feel confident and in charge.

My mind drifted into a meditative state as I fastened my apron and began to chop breakfast veggies. The rhythmic motion of the knife and the satisfying clink of vegetables on the cutting board created a soothing, almost meditative atmosphere.

Going through the motions, allowing my mind to relax for a bit and concentrate on the task at hand, was easy.

Until I was startled out of my reverie by a known voice, I failed to perceive anybody else making their way into the kitchen.

"Wow, you're up, huh?"

Sean was standing at the doorway, resting casually against the frame, when I glanced up unexpectedly. A hint of intrigue shone in his eyes alongside his customary laid-back grin.

"Yeah," I responded, removing an errant hair from my face. Keeping myself occupied is all I'm attempting.

After making his way across the room, Sean reached for the bowl of apples resting on the counter. He took a taste and nodded, clearly comprehending. Regarding that, I see no problem. Maintains mental stimulation.

Because his statements were so honest, I couldn't help but crack a small smile. I was never made to feel insignificant or unwelcome by Sean, one of the few pack members. Just like Molly, he didn't make any assumptions about my omega status or my lack of shift about me; instead, he treated me like any other equal.

"Have you been through anything lately?" As Sean took a seat at the kitchen island, he asked. I've seen a change in you recently. I mean it positively.

I thought about his question for a second before responding. He was correct; I did experience a change. Maybe even lighter. Having a better grasp on my own life.

As I spoke slowly, my hands continued their work, and I guess I've just been trying to figure things out. I will no longer sit around and let other people determine my value. Starting now, I must decide on my own.

An expression of approval spread across Sean's face as he arched an eyebrow. There must be something you're onto here. I finally figured it out on my own after a long period.

His remarks caught me off guard. The confidence and ease with which Sean carried out his duties as Alpha of the pack had always been palpable. Looking at him now, though, I began to think that maybe even the most potent wolves had moments of uncertainty. Perhaps we were all just as helpless as we thought we were.

A ---

We continued to spend the remainder of the morning collaborating in the kitchen as we made breakfast for the pack. I hadn't felt so connected to someone in a long time; I felt like I was a fundamental part of the pack, not merely going through the motions. Sean made me feel at ease just being around him; his conversation was natural and easygoing, and his anecdotes made me laugh, allowing me to temporarily escape the weight that had been pressing down on me.

Upon completion of breakfast, we collectively transported the trays to the dining hall and placed them on the lengthy tables where the group would assemble shortly. I experienced a fleeting sense of fulfillment as the aroma of newly baked bread, eggs, and bacon wafted through the air. I had accomplished this. Nothing spectacular happened, but the fact that it belonged to me made it special.

Sean looked over at me thoughtfully as we finished arranging the table. "Luna, you're stronger than most people give you credit for."

The flattery caught me off guard, and I blinked. I spoke the words, "I'm not sure about that," as my usual skepticism began to take hold.

Sean said, "I am," maintaining a firm but compassionate tone. "You've managed to persevere despite all the challenges you've faced. You have no idea how much strength that takes.

Even after he departed the dining hall, his remarks lingered in my thoughts, and I couldn't help but wonder whether he might have been correct.

A ---

The group started to trickle into the dining hall later in the morning, their chatter making the room throb with energy. While everyone else ate and talked, I sat quietly at the far back, as is my habit. From across the room, Molly smiled and waved at me; in return, I felt a rush of warmth wash over me.

Little by little, things were shifting. It showed in my demeanor and interactions with others as I made my way through the packhouse. No longer was I unseen. I wasn't skulking around, hoping for a change of circumstances. I had joined this group and was starting to feel like I belonged.

However, the same anxiety around Ronan started to resurface as the day progressed. It was almost a relief not to see him all morning. On the other hand, a part of me that was tied to him by the mate bond longed to know his whereabouts and thoughts. No matter how hard I attempted to separate myself from him, the connection remained a continual reminder that I was still deeply connected to him. The bond hummed beneath my skin, reminding me of this.

Afterward, that afternoon, I sensed his presence while tending to the packhouse garden. I could feel his presence before I even looked up; it was like a slight shift in the air. As I turned to look over the garden's perimeter, sure enough, there he was, standing there, watching me.

Ronan remained silent and did not attempt to approach me, although his intense stare was impossible to miss. My body was responding naturally to the connection we shared, and my heart was racing, but I resisted the urge to let it distract me from the work at hand. I refused to allow him to reintegrate me into the state of bewilderment and doubt that had engulfed me for an extended period.

Ronan paused for a second before turning and walking away; I remained standing there, my hands still in the dirt. I forced myself to ignore the familiar tightness in my chest caused by need. I had already decided. For the time being, I was thinking just about myself. I would not have waited for Ronan to reconsider his decision to accept the connection if he was not prepared to do so.

A ---

I sat in my room that night, thinking about what had happened that day. A bizarre whirlwind of emotions raced around me as a result of Sean's words of support, the meaning I had discovered in my job, and Ronan's quiet presence in the garden.

Things were looking up for me. I was constructing my own life, apart from the one in which I had to wait for other people to choose my destiny. However, the connection with Ronan persisted, influencing me in ways I couldn't quite put my finger on. Even when I redirected my attention inward, I couldn't escape the nagging sensation that something was still lacking.

But I was making do for the time being. And perhaps, it was sufficient.

# 13

## Trouble

**LUNA**

The days had started to blend into one another, with a steady cadence of labor, occasional respites with Molly and Sean, and the ever-present, palpable anxiety of Ronan's proximity. I had become accustomed to the constant hum of our mate bond, which persisted even when he wasn't physically present. I didn't think things would remain that way indefinitely. I had a feeling something was going to snap eventually.

I could sense it right now. That change is immutable. The time had come.

A ---

Beginnings were modest. The rogue sightings near the borders were the subject of quiet discussions taking place behind closed doors, and the whispers around the packhouse were getting louder. The tension was tangible, yet nobody knew what had happened. Typically a quiet and collected force within the packhouse, Ronan's warriors were now bustling about as if getting ready for something. As I made my way through the halls, I caught sight of their glum expressions and their hurried, clipped speech.

There was a problem.

Molly entered the kitchen that morning, but her ordinarily upbeat attitude was conspicuously lacking as I kneaded dough for the day's bread.

"Have you caught wind of this?" She glanced around to make sure no one was listening as she asked in a quiet voice.

I took a moment to wipe the flour off my hands. Then what?

"Rogues," she murmured, drawing nearer. Sightings have been reported close to the boundary. They are concerned that a raid might be imminent.

A shiver went through my body. The renegades posed a threat since they lacked allegiance, regulations, and a pack to protect them. They were social outcasts who preyed on smaller groups in search of food. Rogue raids were something I had heard about in the past, but they always felt far away, like they happened to other packs and never to us.

"What does our situation imply?" My voice was barely audible as I whispered the question.

Nervousness visible on Molly's face, she shrugged. "I just don't know. Ronan and his army, however, have been making preparations. "Something big is coming," I thought to myself.

Nodding, my thoughts raced. Even though Ronan's pack had a reputation for brutality in combat, the thought of a renegade raid taking place in our domain was horrifying. Worst of all, I had this nagging suspicion that this wasn't some random attack. A more fundamental factor was involved.

A ---

Tensions in the packhouse rose as the day progressed. Howling and huddling closer together, the wolves discussed potential future events in hushed tones. There was a palpable sense of unease when the usually steady rhythm of the group was upended.

I did my best to occupy my time by attending to my responsibilities in the kitchen and offering assistance wherever I could. I couldn't shake the feeling that things were going to shift, and thoughts of the renegades and Ronan kept popping into my head.

In search of relief from the oppressive atmosphere within the packhouse, I went for a stroll outside in the late afternoon. A grey blanket hung low over the horizon as the sky was gloomy, and the air had the keen bite of impending winter.

I could hear footsteps following close behind me as I made my way into the woods. Sean, who usually sports a carefree look, was jogging to catch up as I turned around to see him.

He drew to a halt next to me and yelled for someone to speak to. ("Are you alright?")

To protect myself from the cold, I wrapped my arms over myself and shrugged. "I am uncertain. Something is... wrong.

As if he could sense the change in the atmosphere, Sean glanced across the tree line and nodded. It's not just you. A storm is coming. Ronan and his men are preparing for something significant, but I have no idea what it is.

A band of outlaws? With the answer already in my head, I asked.

Sean said the word "maybe," but his voice betrayed his lack of certainty. "Perhaps even more."

Even though the thought of "something more" made me shudder, I forced myself to ignore my fears. I had no say over what was about to happen. All I could do was get ready for it.

A ---

The packhouse was a hive of activity first thing in the morning. The main hall was filled with Ronan's troops, who were speaking in low, frantic voices. As always, Alpha Titan commanded attention with his imposing presence and intense gaze at the heart of the gathering. I preferred to remain hidden, keeping my distance so as not to attract any unwanted attention.

However, I was unable to divert my attention from him.

Today, he appeared more focused and distant. Something deeper and sharper had replaced his customary icy detachedness. Even from across the room, I could feel the tension emitted by him in waves, and the mate bond pulsed beneath my skin, bringing back memories of our connection.

While I observed him, Ronan briefly met my sight with a flick of his eyes. Even though it was only a fleeting sight, it caused my heart to race. I had never seen anything like that in his countenance. Perhaps worry? Was it anything else entirely?

Sean showed up by my side, looking downcast before I could think about it. "Seems like it's taking place."

"What is it?" I looked away from Ronan for a second before asking.

Ronan's warriors will meet the renegades, as Sean elucidated. More and more people have reported seeing them, and they are making their way towards the border. There will be no missteps on his part.

I felt a surge of terror as I forced myself to swallow. "Is there a threat to the pack?"

No, Sean shook his head. No, not just yet. Stay away from them, Ronan—he's brilliant at this. We must be prepared in case the situation gets worse.

I forced myself to ignore the overwhelming fear that washed over me at the prospect of Ronan charging into war. His strength was unparalleled by anyone I had encountered. He was the only one who could have dealt with this. Even yet, my heart ached at the thought of him fighting dangerous outlaws as our mate link throbbed.

A ---

After Ronan and his men had departed, the packhouse fell into an eerie silence later that night. As they waited for word, the majority of the pack assembled in the common spaces, where they conversed in low voices. In an effort to divert my attention, I joined Molly, Sean, and a small group of others seated in the room's rear.

However, the constant anxiety in my chest made it difficult for me to concentrate on anything else.

I kept looking out the window, half expecting Ronan and his soldiers to come back at any second while the hours ticked by. Even though the night dragged on, they remained elusive. I was growing increasingly worried as their absence drew on.

Molly whispered, "They'll be fine," when she noticed my restless hands and legs. "Ronan performs like a pro."

I responded with an uncertain "I know," not sure whether I really believed it.

The truth is that we couldn't help but feel a strong attachment due to the mate link. I couldn't resist the attraction I felt for Ronan despite his isolation and his refusal to acknowledge our relationship. My pulse was racing, and my nerves were fraying since the relationship was more vital than ever before as he faced danger out there.

A ---

Eventually, the packhouse doors swung wide in the late afternoon, and Ronan's fighters marched inside. They were filthy, covered in blood, and completely worn out, yet they were still alive. As I looked over the gathering, trying to find Ronan, a wave of relief washed over me.

My heart skipped a beat when I laid eyes on him.

He appeared to be uninjured, yet his expression sent shivers down my spine. A complicated, terrible expression had replaced his customary icy indifference, and it told me the conflict with the outlaws had been more dangerous than I had imagined. From across the room, I could make out a glimmer of something vulnerable and raw in his darker, more distant eyes when they met mine.

However, Ronan abruptly walked away before I could comprehend what he was saying, leaving me to wait there anxiously.

A ---

My thoughts kept returning to the events of the previous day as I lay in bed that night. Too much happened—the renegades, the fight, Ronan's expression upon his return—to comprehend. Before Ronan accepted the bond, I had planned to devote my attention to myself. But now that the threat was accurate and the mating link was pulling me more complicated than before, I wasn't sure.

I couldn't make sense of what was going on; it was too enormous. And it still hadn't ended, whatever it was.

# 14

# Ignition

**LUNA**

A stifling fog hung over the packhouse as the sun rose the following day. What happened the night before, with Ronan's warriors returning wounded but alive and the knowing looks that spoke a great deal about what had happened with the outlaws, lingered with me. Above everything else, though, the expression on Ronan's face stayed with me the longest. I felt unsettled as if we were on the brink of something dangerous because he had undergone a profound and naked transformation.

For the better part of the morning, I stayed in the more peaceful areas of the packhouse, avoiding the busier areas so that I could concentrate. Attempts I made to concentrate on the work at hand were continually interrupted by thoughts of Ronan and the conflict. I couldn't help but be curious about what transpired outside; I wanted to know why he came back looking so downcast.

Someone had to tell me.

A ---

My decision was solidified by noon. I couldn't continue to ignore him. The tie was pulling at me more and more with each passing hour, no matter how much I tried to ignore him, focus on myself, and give him space. The uncertainty became intolerable after last night's encounter when I saw the impact the battle had on him.

We had to have a conversation.

In the training fields, encircled by his warriors, I discovered Ronan. As they moved quickly and precisely through drills, the sound of swords clashing and punching bags being pounded filled the room. Their training was so intense, the implicit urgency in each strike apparent even from a distance. They were getting ready for a monumental event.

As Ronan and his men went through the drills, I stood on the edge of the grounds and watched. His attention was laser-like, and he deliberated and calculated with each step. Unprecedented rigidity in his shoulders and a tense expression on his face betrayed an internal struggle.

Not knowing if it was appropriate to approach him at this moment, I hesitated. Nonetheless, I reached my breaking point. My pulse racing, I inhaled deeply and moved forward.

I yelled out, "Ronan," more loudly than I had planned.

He halted in the middle of his strike, his body immobilizing as he pivoted to confront me. The entire training area went dark as all eyes shifted to me. I tried to push aside the weight of their questions and looks of astonishment and inquiry.

While wiping sweat from his brow, Ronan made an unintelligible gesture for his soldiers to press on without him. He approached me with an expression that was both harsh and distant, yet I could see in his eyes that he wasn't trying to be as unfazed as he seemed.

He asked Luna, "What is it?" in a low but insistent voice.

In an effort to calm myself, I swallowed. "I must speak with you. Concerning the outlaws. In regards to... our own.

His jaw clenched, and I worried he could bolt from the training grounds; nevertheless, he eventually nodded and waved me off. As we strolled side by side, the air was so heavy with tension that neither of us dared to speak. The mate link was a constant reminder of the connection we shared, whether I wanted to admit it or not, and it made my heart race.

Ronan guided me to a remote area of the packhouse grounds close to the forest's border. With their lengthy shadows across the grass, the trees here made for cooler air. With his dark and guarded gaze, he turned to face me and paused.

Enquiring minds need to know. With a raspy voice, he enquired.

Not knowing where to start, I paused for a while. "What transpired during the nite? Among the outlaws?"

Ronan's face became more angular. A struggle ensued. We managed to push them back, but they're reorganizing. We're not done yet.

I gave him a silent nod, even though his response only made my questions worse. And the group? What is the hazard?

A fleeting expression—maybe worry?—crossed Ronan's face as his eyes darted, but he quickly covered it up. "The pack is safe with me. I have to do that.

He spoke with conviction, but I could hear the unease in his voice that suggested he wasn't entirely sure of himself. The renegades posed a significant danger, and I had the impression that this wasn't an innocent assault. More than just a border clash was taking place.

It wasn't, however, my exclusive concern.

"And where do we stand?" My voice was now lower, but my determination was still the same as I asked. "Where do we stand?"

When Ronan locked eyes with mine, his jaw tightened, and his pupils widened. He remained silent for a while, and I could sense the intensity of his inner conflict, the war he was fighting with himself.

"Luna, there is no 'us,'" he finally stated in a controlled and subdued voice. "I've already informed you. No way—"

"You're lying," I cut in, but the words came out faster than I could control them. "You're deceiving yourself. Our connection is genuine, Ronan. No one else feels it the way I do.

I refused to budge, even though his rage was palpable. Just couldn't. Avoid at all costs. His voice was a low growl as he turned away, seemingly trying to finish the debate.

However, I had more to say. "You're scared," I remarked, my voice unwavering even though my heart was racing. "The implications of the link scare you. Think about how much it will drain you. However, evading it won't work indefinitely.

With his back turned and his fists crossed at his sides, Ronan froze. The air was heavy with tension as we sat in quiet contemplation for what seemed like an eternity. The gravity of my statements hung in the air, and their integrity was impossible to deny.

At last, Ronan pivoted around to meet my gaze once again, his countenance resolute, yet his pupils... transformed. Something had changed inside of them—something unfiltered, something untamed.

"You just don't get it," he said, his voice raw with sorrow. I've suffered enough losses. I am unable to risk losing anything else.

I felt my breath escape my body as his words landed like a tonne of bricks. It was one thing to know that Ronan was guarded and had secrets, but to hear him confess it and feel the vulnerability in his voice was a completely different story.

"You won't lose me," I said as I drew nearer. "Ronan, I am your mate. I am not leaving.

After a brief moment of tenderness, Ronan shook his head, and his visage returned to its stony state. "Luna, I am not doing it. The bond has no right to dictate my actions. No way.

The rejection he showed me weighed heavily on my heart. I urgently wanted our talk to make an impact, to help him realize that the link isn't necessarily a weakness. I realized that Ronan wasn't prepared to accept it now that I stood in the shadow of his fear. Perhaps he would never have been.

An emotional "I understand" escaped my lips as I muttered. The problem is that you're incorrect, Ronan. You are not rendered weak by the bond. Strengthens you. Collaborating could make us more powerful.

Ronan remained silent. With his eyes fixed on mine, he remained motionless and silent, his defenses up. At that exact second, I realized that my options had run out. Ronan had taken the decision.

Our connection was unbreakable and powerful, yet it fell short. No, not just yet.

Leaving Ronan to stand alone in the shade of the trees, I turned and walked away with a sorrowful heart.

A ---

Everything that followed was a haze. In an effort to numb the pain in my chest, I dove headfirst into my work. My mind kept playing the conversation with Ronan, and with each iteration, it left me more bewildered and upset.

The link wasn't scary, so why didn't he realize that? We were destined to confront this together; why couldn't he see that?

Sitting outside, I gazed up at the night sky as the sun hollowed below the horizon and the packhouse grew silent. For the first time in a long time, I felt utterly alone as the stars glowed coldly and far above.

It seemed like all my hard work and progress toward finding my niche in this pack were going down the drain. The relationship was something Ronan was opposed to. Sophie continued to bully. A shadow was cast over everything as the rogues' menace grew closer by the day.

To protect myself from the chill that had set in, I embraced myself tightly. The next thing I knew, I couldn't have imagined. I was unsure of how I would deal with it.

One item, though, stood out.

It had been set ablaze.

Plus, it wouldn't go out of power any time soon.

# 15

## Doubt

**LUNA**

The following day came, but the impact of my argument with Ronan the night before still weighed heavily on me. It was heavy and unmovable, and it hung on to me like a shadow. Our connection lingered in the recesses of my consciousness, but it seemed different now—tense, as if a cord were being tugged too taut and about to break.

A feeling of impending terror began to take hold of me. I couldn't put my finger on it, but I could sense an unseen force building beneath the surface. Everything seemed to fit together like a jigsaw puzzle with the stray assaults, Ronan's terror, and the growing tension within the pack.

The more I pondered the matter, the clearer it became that Rogue and Ronan's unwillingness to embrace the mating link weren't the only factors at play.

There was a problem.

A ---

I attempted to occupy myself with the mundane task of making food for the pack, so I spent the majority of the morning in the kitchen once again. Dicing veggies, kneading dough, and whisking ingredients let me concentrate on the work at hand and forget about my worries. However, the increasing anxiety I was experiencing kept resurfacing even as I labored.

Soon after, Molly came over to work side by side with me, and I felt much better in her company. Still, not even Molly was as chatty as usual today. A somber stillness had replaced our customary lightheartedness, and neither of us could break it.

The stillness was broken by the sound of voices rising from the main hall around mid-morning. With our eyes meeting, Molly and I were both immediately on high alert.

Whoa, what's happening? "What am I doing?" I questioned, putting down the carrot-chopping knife.

With a grimace on her face, Molly dried her hands on a towel and answered, "I don't know." "However, the sound is off-putting."

A large number of pack members had already assembled in the main hall, so we both silently proceeded to join them. I forced my way through the crowd in an effort to gain a better view of the action. Standing at the entryway, Ronan's countenance was icy and impenetrable as he spoke with one of his soldiers. I finally made it to the front when I noticed him.

The curiosity of the crowd had not been piqued by Ronan, though. The renegade wolf stood before him.

Upon taking it all in, my heart skipped a beat. The outlaw appeared grimy and battered, with ripped cluthing and a defiant look belying his apparent ailments. Two of Ronan's fighters stood on either side of him, their stances rigid and prepared to move at a second's notice; he wore his hands bound behind his back.

Spotted him skulking close to the southern border," one of the fighters said with a commanding voice. "We've caught him eavesdropping."

As the temperature in the room dropped, Ronan's pupils constricted. "An undercover agent?"

A hint of terror was seen in the rogue's eyes, but he remained motionless as Ronan stared down at him. He spat out, "I'm no spy," in a harsh voice. "I'm merely attempting to stay alive."

The rogue's hold on the warrior's arm was tightened as he was called out in a low growl by one of the fighters. "Our patrols observed you. Apart from espionage, what were you up to?"

The renegade kept silent despite clenching his jaw.

With a step forward, Ronan loomed over the renegade, his presence commanding. Why are you on my land if you aren't eavesdropping?

As the outlaw fixed his stern gaze on Ronan, a prolonged, nervous hush fell over the room. I was afraid he wouldn't respond at first, but he eventually did, speaking in a low, poisonous voice.

"I'm not the sole one," he declared. Others exist as well. And we will pursue you.

With each passing whisper, the air in the room became increasingly thick with anxiety. As I listened to the renegade speak, my chest tightened with fear. This wasn't an arbitrary assault. Something was organized and arranged for this.

Ronan's attitude turned icy and calculated as his eyes grew darker. "Do you really believe you have the power to intimidate me?" he inquired, his voice audibly nervous.

An evildoer scowled. "There's no danger here. It's an assurance.

The audience went silent for an instant as the rogue's words hung heavy in the air, threatening to explode at any second. Anxieties were building in the audience, and I could sense that everyone was wondering, "How many rogues were there?" Just how near were they? Did we face a more significant threat than we had anticipated?

With a determined look on his face, Ronan turned to his army. Bring him to the prison. Later on, I will handle him.

The warriors dragged away the rogue as the audience started to disperse slowly, and they nodded in agreement. However, the atmosphere remained charged with anxiety. It actually got heavier. Words spoken by the renegade would remain long after they had been spoken.

A ---

Feeling anxious and unsettled, the remainder of the day flew by. Whispers circulated among the pack members as they worried about the rogue's danger and its potential consequences. Wolves discussing things like how many renegades were out there and whether or not we were ready for another assault were overheard as I made my way through the packhouse.

More than that, though, I had the uneasy sensation that underlying forces were at work. It wasn't only the outlaws that were involved. Something else was happening, and I wasn't sure what it was.

As night fell, I found myself aimlessly roaming the packhouse, my thoughts consumed by the events of the previous day. My anxiety level skyrocketed as the rogue's remarks resounded in my mind. Is it possible that this is just the start? Were we to underestimate the extent to which the pack was threatened?

Next, there was Ronan.

The animosity between us continued to linger in the background of my mind, and I hadn't talked to him since our disagreement the day before. I sensed his concern for the pack and the rogue assaults, but I couldn't help but wonder if there was something crucial he was withholding from me.

A ---

My thoughts kept returning to the events of the previous day as I lay in bed that night. Everything was making me nervous: the renegade's stubbornness, Ronan's emotionless reaction, and the mounting terror that gripped the pack.

Things seemed to be getting out of hand, and I had the uneasy sensation that something terrible was about to happen. I took the renegade's word for it that more were on the way. I couldn't fathom the rationale behind it. Why now? What gives?

Also, why did Ronan seem to know more than he was letting on to me?

Underneath my skin, the mating link pulsed subtly, serving as a continual reminder of our connection; nonetheless, it felt distant and frail tonight. I was cut off from him and had no idea what was going through his mind. One thing was sure, though: the rogues weren't the only ones involved in whatever was going on.

I sensed an impending event.

I also doubted that we were prepared.

A ---

This chapter furthers Luna's notion that there is something more going on behind the scenes and introduces the possibility of a more comprehensive rogue attack. As they get ready for what's to come, the animosity between the pack is getting worse, and Luna isn't sure what part Ronan will play. If you would like to proceed or request changes, please inform me.

# Epilogue

## A Storm on the Horizon

As the sun was hollow below the horizon, Luna stood on a hilltop looking out over the pack's domain and the woodland below. Even though she had to fight for her position in the pack, she finally felt at home there a few months after her return. Standing silently by her side, Ronan and her link were more substantial than before, forged afresh by the hardships they had endured and the strength they had developed as a team.

However, the forest had a distinct vibe this evening.

A heavy calm permeated the air as if the trees were holding their breath, as the customary harmony of birdsong and rustling leaves ceased to exist. An unsettling sensation that Luna hadn't experienced since her days of fleeing—being pursued—came creeping up her spine as her senses became acute. But she had transformed from a lone wolf. Something dark and terrifying was approaching, and now she was a part of something bigger.

"Is it physical?" Ronan's posture conveyed a great deal of emotion, even though his voice was hardly audible. He stood stiffly, his gaze locked on the faraway hills.

Nodding, Luna's limbs tensed, and her heart raced with an unexplainable sense of anticipation. "Something seems to be on the horizon. Something is keeping tabs on us.

After a little pause, Ronan looked down at the pack members, who were smiling and feeling good about themselves. Even though nobody else saw the shift, Luna could see Ronan's expression change as she felt a similar stirring in her own heart.

There was going to be a test of the peace they had worked for.

Her mind was racing with possibilities as she questioned, "Who do you think it is?" She was trying to keep her voice steady. Those pesky old foes? Enemy packs far away? (Alternatively, it could be something more sinister, a specter from her history that she had hoped to put to rest?).

Returning his gaze to her, Ronan clinched his jaw and widened his pupils. "I am uncertain. Still, we must be prepared.

She felt a rush of pressure. Her hard work, her restored family, and the life she had started to rebuild were all in jeopardy. Luna sensed that the challenges they had encountered were but the start as the sun dipped below the horizon, spreading lengthy, foreboding shadows across the woodland.

A chilling, bitter aroma that had been absent for years was carried on the wind, making her skin crawl. As they hunkered down, ready for anything, she sensed Ronan's grip on hers become more firm.

The far mountains resounded with the first rumble of thunder.
They could no longer see the storm on the horizon; it was barreling down on them.

## *About* the *Author*

**ADURA GBEMI ADEGBOYE** is an African manager, business administrator, entrepreneur, and motivational speaker. ADEGBOYE has his BA from Yale University, IPMA from Adonai University, and a Masters in Business Administration (MBA) from the University of Salford, Manchester. He was born in South Africa but is presently based in Nigeria as a motivational speaker in institutions, sectors, and seminars with young and upcoming managers all over Africa.

# Acknowledgments

First of all, To God be the glory for completing this medical write-up. It could have been impossible without the help of God, a present help and sustainer whose wisdom made this medical manuscript possible and successful.

My appreciation also goes to my family, friends, well-wishers, and many others I cannot start mentioning, who have always been there in one way or another from the beginning of this manuscript till the end of it.